GROWERS HANDBOOK SERIES
VOLUME I

SEED PROPAGATED GERANIUMS

The why and hows
of production of hybrid geraniums

Allan M. Armitage

Department of Horticulture
University of Georgia
Athens, GA 30602 U.S.A.

TIMBER PRESS
Portland, Oregon

© Timber Press 1985
All rights reserved.
Printed in the United States of America.
ISBN 0-88192-064-9

TIMBER PRESS
9999 SW Wilshire
Portland, Oregon 97225

Contents

Introduction   1

Cultivars   2

Culture   3

      Factors Affecting Flowering   12

      Physiological Disorders   20

      Timing and Scheduling   24

Pests and Diseases   30

Costs of Growing Hybrid Geraniums   32

Future   36

Appendix   41

References   42

Introduction

This manual puts in print information concerning the seed propagated geranium—a crop which is but a mere baby compared with its older sister, the vegetatively propagated zonal geranium. In less than 20 years the hybrid geranium, as the seed propagated geranium has come to be known, has come from nowhere to become a significant part of the bedding and potted plant industry. Since working with my first 'Carefree' cultivar many years ago, I have become fascinated with the means of influencing flowering and maintaining quality of this crop. I also wish to point out that although this manual restricts itself to the hybrid geranium, much of the information is applicable to all geraniums. I have written this manual, not because I believe that hybrid geraniums have any particular advantage over vegetative geraniums or vice versa, but simply to provide growers with sound information on growing hybrid geraniums successfully. I believe that geraniums, whether they be seed or vegetatively propagated, will continue to be popular and that growers will choose to produce the type of crop with which they are most comfortable based on marketing and management decisions.

The production of hybrid geraniums has branched out into four production methods. The first method is what I term the "classical" pack method where the grower sows in seed flats, transplants into final containers and produces the crop from beginning to end himself. This method allows total control of the crop without the outlay of funds to purchase seeding machines and other "plug" related equipment. This has been a method used for years by bedding plant growers but is on the decline. The advent of the single cell transplant or "plug" has made significant change in the growing of geraniums from seed and has produced the specialist plug producer. These growers grow the plant in a plug tray containing anywhere from 200 to 600 cells for 6–12 weeks whereupon the green plants are shipped to growers for finishing. This brings us to the third kind of geranium producer, the plug finisher. His main job is transplant the plug to its final container and put a flower on the plant. The fourth method is for growers who produce plugs for their own finishing. This has become a very common means of production of hybrid geraniums and also allows total crop control.

This manual is intended to be of use to both the beginning grower as well as veteran growers and researchers. The information is pertinent to both plug and pack growers and where major differences exist between the production techniques of the two, they will be pointed out. It is as up-to-date as possible and although it will not answer all questions on hybrid geranium production, I hope it will satisfy the hunger of some and provide food for thought for others.

Cultivars

Since the development of 'Nittany Red Lion' by Drs. Darrell Walker and Dick Craig at Pennsylvania State University, the search for cultivars with unique colors, growth habits and flower shape has increased internationally. In the late 1960's, the 'New Era' and 'Carefree' series emerged. There were a number of problems with the early cultivated varieties. They weren't very floriferous, they didn't last long and they all took a very long time to bloom. Sowing in December could produce a flowering crop in May but then again so could a January sowing. Grower resistance also was high to seed propagated geraniums with their single flowers which shattered readily. Since then the effort to breed early flowering cultivars in order to reduce production time has been one of the constants in geranium breeding. In recent years seed firms have produced an incredible number of new cultivars combining many different plant characteristics and exhibiting all colors except blue and yellow. Today there are a wide array of cultivars from which the grower can choose depending upon his marketing objectives—tall or dwarf, double or single flowers, and differing zonal qualities.

Some of the better known groups of cultivars grown include the 'Carefree' series, the 'Sprinter' series, the 'Ringo' series, the 'Flash' series and the 'Orbit' series. Each series consists of plants of similar habit and usually includes flower colors of white, red and salmon. Many of the more popular cultivars grown come from these series but others include 'Mustang', 'Red Elite'. 'Sooner Red', 'Cameo' and many, many more. Each cultivar has unique characterstics but as more and more come on the market, the differences between them become less dramatic. Information concerning the height of the plant at maturity, the size of the flower and flower head as well as the time to flower from sowing can be found for most cultivars on the market today (6,7,9,13,31). There are pack and garden trials in the flower breeder's locale, at the major seed firms and at certain universities every year. Table 1 lists some general information on many but certainly not all of the cultivars produced in the last 5 years. More detailed information on most available cultivars can be found from the various breeders and in the bibliography.

Table 1.	Flower color, pack height, and flowering time of 74 cultivars of hybrid geraniums grown from 1979–1984.		
Cultivar	Flower Color	Height[z] in Pack	Flowering time[y] in Pack
Accent	Red	MT	M
Adretta	Scarlet	MS	M
Amaretto	Salmon	S	M
Applause	Salmon	MT	M
Appleblossom Orbit	Salmon-Pink	MT	M
Appleblossom Orbit Imp	Salmon-Pink	MT	ME
Bright Eyes	Bi-color	MT	M
Cameo	White	MT	M
Century Cardinal	Brick-red	MT	M

2

Cultivar	Flower Color	Height[z] in Pack	Flowering time[y] in Pack
Cherie	Salmon	MT	L
Cherie 80	Salmon	MS	M
Cherry Diamond	Carmine-Red	MS	E
Cherry Glow	Carmine-Red	MS	M
Coral Orbit	Coral	MS	ME
Delta Queen	Orange-Red	T	M
Encounter Red	Red	MT	M
Encounter Salmon	Salmon	MS	E
Fireflash	Red	MT	M
Gremlin Coral	Coral	S	M
Gremlin Red	Scarlet	S	M
Heidi	Bi-color	MT	L
Hollywood Red	Red	MT	M
Hollywood Salmon	Salmon	MT	M
Hollywood Star	Bi-color	MS	ME
Hollywood White	White	MS	ME
Ice Queen	White	MT	M
Jackpot	Orange-Scarlet	MT	M
Knockout	Red	MT	M
Marathon Dbl Red	Salmon-Red	T	M
Merlin	Violet	MT	M
Mustang	Scarlet-Red	MT	M
Orchid Orbit	Lavender-Pink	MT	M
Picasso	Fuchsia	MT	M
Pink Orbit	Pink	MT	M
Pinwheel Salmon	Salmon	MT	E
Pinwheel Scarlet	Scarlet	MS	M
Playboy Cerise	Cherry-Red	S	M
Playboy Salmon	Salmon	S	M
Quix	Red	MS	M
Razzmatazz	Bi-color	MT	ME
Redhead	Red	S	M
Red Elite	Red	MS	E
Red Express	Red	MS	ME
Red Fountain	Red	MS	M
Red Orbit	Red	MT	ML
Red Pimpernell	Carmine-Red	MT	ML
Red Standard	Red	MS	M
Ringleader Pink	Pink	MS	M
Ringleader Red	Red	MS	M
Ringleader Salmon	Salmon	MS	M
Ringo Dolly	Bi-color	MS	ME
Ringo Rose	Lilac-Rose	S	ME
Ringo Rouge	Carmine-Red	MS	M
Ringo Salmon	Salmon	MS	ME
Ringo Scarlet	Scarlet	MT	ME

S = small   E = early
MS = moderately small   ME = moderately early
MT = moderately tall   M = medium
T = tall   L = late
— = basket type

Cultivar	Flower Color	Height[z] in Pack	Flowering time[y] in Pack
Rose Diamond	Rose-Pink	MT	E
Rosita	Rose-Pink	MT	M
Salmon Flash	Salmon	MS	M
Salmon Orbit	Salmon	MT	M
Scarlet Diamond	Scarlet	MS	E
Scarlet Eye Orbit	Bi-color	MT	M
Scarlet Orbit Imp	Scarlet	MT	M
Sitta	Red	MS	E
Smash Hit	Scarlet-Red	MT	E
Smash Hit Burgandy	Fuchsia	M	M
Smash Hit New Dawn	Coral	M	M
Smash Hit Rose Pink	Rose-Pink	MT	M
Smash Hit Salmon	Salmon	MT	M
Snowdon	White	MT	ME
Sooner Br. Pink	Bright Pink	MT	ME
Sooner Dp. Salmon	Deep Salmon	MS	M
Sooner Red	Red	MS	E
Sprinter Salmon	Salmon	MS	M
Sprinter Scarlet	Scarlet	MT	M
Sprinter White	White	MT	M
Summer Showers	Mix	—	L
Sunbird	Scarlet-Red	S	M
Tara	Salmon-Red	MT	L
Voodoo	Orange-Red	MT	L
White Orbit	White	MS	M

Culture

GERMINATION

All geranium seed available to commercial growers from reputable seed houses is well-cleaned and scarified (usually by acid) to insure rapid germination. Germination of fresh hybrid geranium seed is usually greater than 80% for most cultivars and 90–95% germination is not uncommon when the optimum environment is provided (6,7).

Seed can also be obtained in pelletized form, developed for mechanical seeders to assist in more uniform distribution of seed in the seed tray. They are pelletized with an inert material, usually of clay or vermiculite. Pelletized seed require 2–4 days of additional germination time and are approximately 15–30% more costly (29). With the advent of better methods for cleaning seed, seedsmen are offering "Hi-Tech" or "Hi-Germ" seed which guarantee a high percentage germination rate and a uniform stand. This trend is certainly going to continue as high technology growing methods demand high technology seed.

Uniform germination is essential in any seed propagated crop and hybrid geraniums are no exception. Although air temperature may fluctuate, uniform media temperature of 22–25°C (72–77°F) should be maintained for optimal results. Temperatures below 20°C (68°F) and above 30°C (86°F) result in poorer germination and poor uniformity. Light intensity and duration have little effect on germination but both the percent germination and the time for germination are significantly affected by temperature (Table 2).

Table 2. **Effect of temperature on seed germination of hybrid geraniums.**

	Night Temperature[z]			
	10°C 50°F	15°C 60°F	21°C 70°F	26°C 80°F
Germination percent	80	86	91	89
Days required for germination	16	13	9	7

[z]Propagated under intermittent mist and greenhouse conditions. Adapted from (42).

The germination medium should be of relatively good porosity, well aerated and free of large particles under which seeds could be trapped. The medium should be sterile and low in or totally without fertilizer. The best media for germination are sand, perlite, vermiculite, peat-vermiculite or mixes of these ingredients (24,25). Seed flats or plug trays are recommended for germinating geranium seeds.

Geranium seeds take up water rapidly so adequate water must be applied to the medium to allow the seeds to swell. The medium should be constantly moist but not dripping with free water. In the germination area, a misting system on a solar switch or evaporative leaf mechanism will provide the humidity and free water necessary for uniform germination. Humidity tents have also been used successfully in the past and are regaining popularity once again. The use of timers to activate mist systems should be avoided as they do not take into account the day to day changes in temperature and light. Once the seed starts to swell and the radicle (i.e. root tip) emerges, the germination process is irreversible. Water must not be neglected at this time or the seed will die and germination percentages will be frustratingly low. No malfunction of equipment can be tolerated at this stage.

More and more growers today are using some form of controlled environment germination chambers. These are almost a must for "plug" growers where uniformity and high germination is essential. These rooms maintain a uniform temperature of 72–77°F and a constant humidity of approximately 95%. Although light is often not necessary for germination, light must be provided when the seedlings emerge. Therefore, germination rooms should also provide at least 200–250 ft candles of light from cool-white fluorescent lamps.

Under these conditions seedlings stretch very rapidly and must be removed to a shaded area of the greenhouse within 3–5 days of germination. Germination rooms make a great deal of sense regardless of the method of production (seed flat or plug tray) and help to insure maximum germination with maximum uniformity. They are a wise investment for such an important part of the production cycle.

Growing on plugs: At an informal meeting of researchers and bedding plant growers held in Virginia Beach in 1984, the discussion centered around the differences between handling seedlings transplanted from a seed flat and handling seedlings in plug trays. The common view was that were really no major differences between the systems but that when growing on plugs, one had to *pay much closer attention to detail.*

When bringing seedlings out of the germination chambers, they must be placed out of full sun and misted occasionally for the first day to acclimate to their environment. Seedlings require a temperature of 70–75°F after coming out of the germination chamber (37). The major concern in plug production is that of proper moisture control. This requires a medium which is porous and well aerated to allow for good drainage. Due to the low medium volume, water must be applied more often to plug cells than to pack cells. Producing high quality plugs requires that other environmental inputs such as light, temp, and CO_2 are similar to that of pack or pot culture and done at the proper time. This does not mean to imply that one can be sloppy or lackadaisical with pack culture, however, due to greater soil volumes, pack culture is a little more forgiving than plug culture.

Transplanting from seed flats: Seed flats are generally shallow and contain only a small reservoir of nutrients, if any, so prompt transplanting is necessary. Geraniums may be transplanted within 10 days from sowing. At this stage only the cotyledon leaves have emerged and the root system is small. Obviously significant damage can result if the seedlings are handled roughly. Transplanting can be deferred until 1–2 true leaves develop (approx. 14–21 days) without any significant delay in flowering time but no longer. Experience with other annual plants indicates that a 2–3 week delay of transplanting significantly delays flowering time (11) and geraniums respond similarly.

GROWING MEDIA

Geraniums must be grown in a medium with good drainage and aeration regardless of whether they are being produced in flats or plugs. The great majority of growers use soilless mix which they mix themselves or purchase. Commercially available soilless mixes include mixtures of Canadian peat, hypnum peat, horticulture vermiculite, perlite, or bark mixed in various ratios and may include other ingredients such as sawdust, styrofoam beads and treated sewage. Most commercial mixes also contain some granular fertilizers, such as calcium nitrate, potassium nitrate, superphosphate, iron sulphate, trace elements as well as a wetting agent (Table 3).

Table 3.	Major and minor ingredients incorporated in most soilless mixes used for hybrid geraniums on a liquid feed program.	
Major ingredients		*Amount/yd^3 of mix*
Sphagnum peat		13 bushels
Horticultural vermiculte		13 bushels

Composted softwood bark, horticultural perlite and styrofoam have also been used as major ingredients as a partial substitute for one of the above ingredients.

Minor ingredients	*Amount/yd^3*	
Ground dalomitic limestone	5 −10	lbs
Superphosphate or	1 −2	lbs
Treble superphosphate	0.5–1	lbs
Calcium or potassium nitrate	1 −2	lbs
Trace elements	a	
Wetting elements	3 oz	

a – 3 oz FTE 503, 4 oz FTE 555, 4 lbs Esmigran or 4 lbs PERK per yd^3.

Growers making up their own media must take care to use ingredients in the mix that are useful as well as available. There seem to be as many recipes for soilless mixes as there are "cooks" but many geranium growers have been successful with various "Cornell" mixes (2). Many of today's "bagged" mixes are variations on the Cornell mix formulations. A great many growers use softwood barks as the main ingredient in their soilless recipe because drainage is usually excellent. Recently there has been some interest in hypnum peat as a supplement to or a replacement for sphagnum moss. Hypnum peat is a lake bottom moss and so is widely available. Because it does not have surface waxes, it does not require a wetting agent and re-wets easily. Its pH is between 5.0 and 5.5 compared to 3.0–4.0 for sphagnum, and less lime is needed. This moss may have some potential in the near future.

When growing in plug trays, there is such a small depth and volume of medium that drainage is the most important consideration in a mix. Air porosity should range between 15–18% for optimal growth. See Appendix #1 for method of determining % porosity and water-holding capacity.

It is not necessary to grow in soilless medium but these mixes have become popular because good top soil is often unavailable or too costly. Geraniums have been grown in a mixture of 1 part loam, 1 part vermiculite or perlite and 1 part sphagnum peat moss for many years with excellent results. If good quality soil is available, is consistent in its properties and is inexpensive, then there is no reason that soil should not be used. Muck soils have been used in the past, but are unsatisfactory for bedding plant or pot geraniums. In general, there is no one "best" growing medium for geraniums. If you feel you are not obtaining the results you should, work with your suppliers. They are usually very knowledgeable and always helpful.

WATERING

When young, geraniums are very susceptible to many disease organisms, particularly water molds such as *Pythium* and *Phytophthora*. Young geraniums should be watered sparingly and the media must be allowed to dry out between waterings. More geraniums are killed by overwatering when young than die for any other reason. As the plants increase in size, the watering schedule should be increased to adjust for the extra water uptake by the plant. Geraniums don't physically wilt like most other bedding plants. That is, the leaves don't collapse around the stem like impatiens but remain somewhat turgid. Geraniums which are allowed to dry out however, still undergo adverse physiological changes. Photosynthesis is markedly reduced and they never fully recover their original rate of growth.

Plants grown in 4″ pots which were allowed to dry out took 3 full days to regain maximum photosynthesis after being rewatered and the new photosynthetic rate was less than it was prior to drying out (Figure 1). This means that extra time is needed on the bench in order to grow a plant to market size if water stress has occurred.

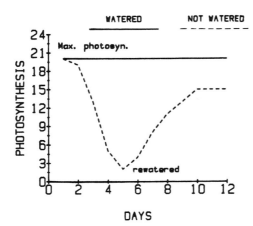

Fig. 1. When plants are not watered, they never recover their full photosynthetic rate. Adapted from (15).

Greenhouse water should be tested regularly for soluble salts, pH, and bicarbonate levels. The pH of the water should be between 4.5–6.5. Water of high pH should be treated through an injector with phosphoric or sulphuric acid to reduce the pH to acceptable levels. Sulphuric acid is gaining popularity as a water amendment because phosphoric acid can tie up iron in the medium and the plants can become chlorotic. The amount of acid added to the water depends on the pH, the bicarbonate level and the injector system so it must be worked out by experimentation.

The amount of water applied during production also affects plant growth. Plants grown with high moisture levels are much taller and grow more than those receiving very little water (Table 4). However, more does not necessarily mean better as high moisture levels often result in poor quality plants with a poorer shelf life.

8

Table 4.	Effect of 3 moisture levels on growth of Geranium 'Dark Red Irene'. From (40).		
Soil moisture regime	Leaf area (cm)	Dry Wt (g)	Increase in height (cm) from from beginning of exp't
High	1589	31.4	19.9
Medium	1167	17.8	13.6
Low	724	13.4	7.5

There is a fine line between too much and too little water with all plants . Plants grown in flats or plugs dry out more rapidly, particularly around the edges, compared with those in 4" or 5" pots. The most important element in watering is not sophisticated equipment or complex schedules, but rather careful and experienced attention by the person who runs the equipment or who holds the water hose. Proper watering is the most important element in the growing of geraniums.

FERTILIZATION

Soluble fertilizers are most commonly used. Growers are rightfully very cautious about applying nitrogen when plants are young. Geraniums, however, are heavy feeders and applying 150 ppm N from emergence does no damage. This is a useful practice when more growth is desired, but feeding too early may not be beneficial when short compact plants are needed as is often the case in plugs. In plug trays, fertilization should be started after 1 week of bench time and then applied at 150–200 ppm N. However, immediate application of 150 ppm N results in more growth than waiting until plants have been "established" before applying a full nitrogen (i.e. 300 ppm N) concentration. More flower buds and more leaf surface area are produced using 150 ppm N at transplant compared with no fertilizer (Table 5). Using 300 ppm N immediately upon transplanting results in substantial injury. Once established, geraniums should be fed with 300 ppm N until color is noticed in the flower buds. At that point, fertilization should be reduced or terminated as continued application will result in reduced shelf life of the plants if they are placed in a low-light setting. Generally, tap water (i.e. fertilizer free water) should be applied at least once every 4 waterings to reduce soluble salt concentrations in the media.

Table 5.	Effects of early nitrogen application to 'Red Orbit' geranium. Sown in November. Adapted from (17).			
Nitrogen conc. (ppm) from transplant to 4 weeks	Days to flower	Leaf surface area (cm²)	Flower buds	Injury (%)
0	113 b[z]	613 b	52 b	0
150	118 a	875 a	65 a	0
300	120 a	640 b	42 c	30

[z]Numbers followed by the same letter are not significantly different using Duncan's Multiple Range Test (5%).

The use of soluble fertilizers high in the ammoniacal form of nitrogen should be avoided for hybrid geraniums during the cool, dark months of December to February. Nitrogen in the nitrate form is more readily used by the plant during this period. Flowering can be delayed if high levels of ammonia nitrogen are applied. The ratio of ammoniacal to nitrate form is found on the fertilizer bags and in greenhouse supply catalogues.

Between the 8–10th week, hybrid geraniums growing in soilless media may appear chlorotic and will benefit from an application of iron chelate (1 tsp/gal, 0.4 g/l). This should be the only micronutrient application necessary if granular trace elements were originally added to the media (See section on soil). Iron chelate may also be added to the dry media prior to planting at the rate of 0.75 oz/yd^3 (40 g/m^3).

Fertilizers high in phosphorus (eg. 20-20-20) are often used but phosphorus is seldom needed in such quantities if superphosphate has been added to the medium. Although phosphorus leaches out more rapidly in soilless mixes than in mineral soils, it remains available much longer than nitrogen and potassium. Heavy feeding with phosphorus is not only unnecessary, but results in hard earned dollars going down the drain.

Granular fertilizers are used successfully with many pot crops including geranium. Osmocote 14-14-14 may be incorporated at the rate of 1 tsp per 6 in. pot. Other granular fertilizers (such as Mag Amp) have been used either by themselves or in combination with a liquid feed program. Many of the granular fertilizers are high in ammoniacal nitrogen so should only be used as a supplement to liquid feed during winter and hot summer months. New products are now being tested with higher ratios of nitrate to ammonia nitrogen. Chlorosis problems may still occur and an iron supplement may be required.

GROWTH REGULATORS

Growth regulators have been used to control the height of hybrid geraniums for many years. They make plants more compact, increase uniformity and accelerate flowering. The major growth regulator used is Cycocel (chlormequat, CCC). Its ability to keep the plant compact regardless of season compensates for reduced light intensity in the winter (51). It is usually applied as a spray, but may be applied as a drench. Cycocel should be applied twice as a spray or once as a drench to most cultivars but many new dwarf cultivars do not require any while some vigorous cultivars require more than two.

For classical production Cycocel is first applied at the rate of 1500 ppm, 5 weeks from sowing and the second one week later. The timing of the first application can be determined by the presence of 3–4 true leaves, the largest being about the size of a quarter.

In the case of plug culture, 750 ppm CCC should be applied when plants have 2–3 leaves and the leaves are the size of a U.S. ten cent piece (approximately 14 days from sowing) and followed by a 1500–2500 ppm application at 28 days. During the summer or in times of high temperatures an additional application may be made at 35 days. When finished plants are grown from plugs, generally one application of 1500 ppm Cycocel is applied approximately 1 week after transplanting to the final container.

Applying Cycocel at 3000 ppm once only in order to save time should be avoided as leaf burn can be quite serious on some cultivars. One gallon of diluted growth regulator should cover approximately 100 flats or 1200 4" pots.

It should be applied the morning of a sunny day, if possible, to dry the growth regulator quickly. If applied in the heat of the day (i.e. >80°F) or on a cool, cloudy day when drying is slow, some marginal leaf damage and yellowing may result (Fig. 2, back cover). In general the yellowing will subside and not be visible when plants are marketed.

The growth regulator A-Rest (Ancymidol) is also effective on geraniums at the rate of 200 ppm applied either as a foliar spray or a soil drench. When used on plugs it should be applied at ½ dosage at 14–20 days followed by a second application (½ dosage) 7–10 days later. Both Cycocel and A-Rest provide the added benefit of inducing flowering approximately 1 week sooner when applied early in the life cycle of the plant (41). Some producers use several applications of growth regulators at lower concentrations to reduce marginal burn while obtaining control of plant height. Applications as low as 350 ppm cycocel applied 6–7 times have been used. Care must be taken, however, that the concentration used is high enough to retard growth. It matters little if a grower applies a growth regulator 10 times if the concentration rate is too low to affect the plant. See Table 6 for mixing ratios.

Too much growth regulator results in excessively small plants (Fig. 3, back cover) which if not fit for the novelty store are suitable only for the garbage pail.

Table 6.			Mixing growth regulator solutions.		
CYCOCEL	11.8% active ingredient				
Spray	Spray solution		fl oz of Cycocel/ gallon of final solution		ml of Cycocel/ liter of final solution
	1500 ppm*		1.6		12.7
	2000 ppm		2.2		16.9
	3000 ppm**		3.2		25.4
Drench	Dose (mg/pot)	Drench vol per 6" pot	ppm	fl oz of Cycocel/ gal of final sol.	ml of Cycocel/ liter of final sol.
	532.3	6 fl oz	3000	3.254	25.42

*Commonly called 1:80.   **Commonly called 1:40.

A-REST	0.0264% active ingredient				
Spray	fl oz of A-Rest/ Spray Solution		ml of A-Rest/ gallon of final solution		liter of final solution
	10 ppm		4.8		37.7
	50 ppm		24.2		189.4
	100 ppm		48.5		378.8
Drench	Dose (mg/pot)	Drench vol per pot	ppm	fl oz of A-Rest/ gal of final sol.	ml of A-Rest/ liter of final sol.
	0.25	8 fl oz	1.1	0.5	4.0
	0.75	8 fl oz	3.2	1.5	12.0
	1.00	8 fl oz	4.2	2.0	16.0

Some new materials are presently working their way into the marketplace such as PP 333 (paclobutrazol, Bonzi) which appear to have potential as a growth retardant for geraniums. Preliminary research at the University of Georgia indicates that a first application of 2.5 oz/gal (78 ppm) when the plants are approximately the size of a U.S. 50 fifty cent piece followed by a 2nd application 10 days later results in excellent height control.

Other growth regulators such as ethyl (ethrephon) and gibberelic acid are used to increase cutting production on stock plants of zonal geraniums but are seldom used on seed propagated types.

Factors Affecting Flowering

LIGHT

The main factor affecting time to flower in hybrid geraniums is light inten-sity. Geraniums are day-neutral plants, thus day-length is important only because plants are receiving light for a longer or shorter period of time. Geraniums are most responsive to changes in light intensity when young. So, greenhouse covers should be clean and free of shading within 1–2 weeks after transplanting.

Flower initiation is directly influenced by the cumulative light the plant receives to that time (28). Time of initiation varies depending on cultivar and geographic location but usually occurs between 56 days and 80 days from sowing under ambient conditions. The inherent potential for earlier initiation has probably not yet been realized and flowers have been initiated experimentally in less than 5 weeks.

Flowering is most directly influenced in the first 6–8 leaf stage of geranium growth. These leaves must be healthy and exposed to as much light as possible as it is the photosynthetic activity of these leaves which provide carbon-containing substances (sugars and starches) to the developing meristem (18). While it is true that the total leaf area of shade-grown plants is greater than that of sun-grown plants at time of flowering, the initial 6–8 leaves are produced sooner and are larger when grown in high light compared with shaded conditions (See Fig. 4). It takes more time therefore for the meristem to obtain the substances to initiate reproductive growth when light is low.

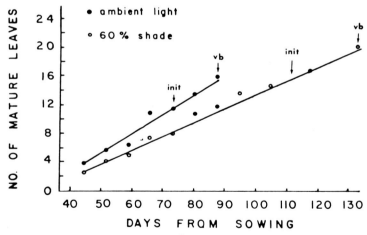

Fig. 4. Mature leaves are developed later on plants grown under shaded conditions. Init = flower initiation; vb = visible bud. Adapted from (19).

High light intensity accelerates the development of the plant from a leaf producer (i.e. vegetative state) to a flower producer (i.e. reproductive state). I have classified the life cycle of a geranium plant into various stages (Fig. 5). Although on paper the distinct stages are sharply delineated, the end of one stage and the beginning of the next are usually rather indistinct. Such is science!

Light intensity greatly affects the vegetative growth phase of the plants. That is, less time is required in the vegetative stage when light levels are high.

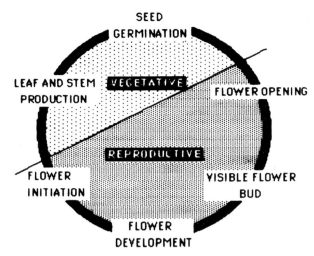

Fig. 5. Developmental phases of the geranium.

Flower development is the time when the flower parts are made—sepals, followed by petals, stamens and finally the 5-lobed pistil, all formed with precision and grace. The formation of these parts is also accelerated when plants receive high light levels.

Once the bud is visible, the role of light is that of maintaining the quality of the buds (i.e. bud number, bud size), but has little to do with the time required for further development. In fact, if a budded geranium is placed in near-darkness, the flower will open at the same time as the flowers on plants which remained on the bench, but the flowers will be of poor quality. The moral here is that once visible bud stage has been reached, regulating light intensity is not an effective way to accelerate or slow down the crop. At this stage, temperature becomes the major controlling factor (see section on temperature).

Although this section deals with the effects of light on flowering, there are other obvious effects of light on growth as well. Low light reduces dry weight, causes thin leaves and results in a reduction of sugar and starch content of leaves regardless of the temperature at which the plants are grown (51).

Supplemental Light: Supplemental lighting of hybrid geraniums significantly reduces the time to flower, ensures sturdy, compact seedling growth, early leaf expansion, and increased branching. Although many types of light units exist, the main sources of supplemental light are cool-white fluorescent, high-pressure sodium lamps (HPS) or metal halide lamps (MH). The HPS and MH lamps are better as supplemental sources than fluorescent tubes because they block less natural light and are more efficient in their utilization of electrical power. Low pressure sodium (LPS) lamps should not be used.

Plants should be lit when young even in the seed or plug flat and in the initial weeks in the transplant tray. Although some growers light for 24 hours a day, 18 hours per day up to 6 weeks appears to be satisfactory. Lighting for more than 6 weeks does not provide enough benefit to warrant the additional expense (5, 27). Light intensities used commercially are between 250–500 f.c. but higher light levels induce faster flowering. Continuous high light intensities of 1200 f.c. result in flowering plants approximately 85 days from sowing (See Fig. 6.). Incidentally, the least amount of time recorded from sowing to flower is 29 days for 'Cherry Diamond' (22).

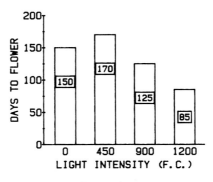

Figure 6

Supplemental light in combination with Cycocel accelerates flowering in an additive fashion, that is the plants flower faster with both supplemental light and Cycocel than with supplemental light or Cycocel alone (Fig. 7). For reasons mentioned previously, lighting after the visible bud stage in an effort to accelerate flowering will do little except provide some extra heat and increase expense.

Lighting of geraniums also allows the plant to better use extra carbon dioxide, and supplemental winter lighting and supplemental carbon dioxide are natural partners for more rapid production of geraniums (see next section).

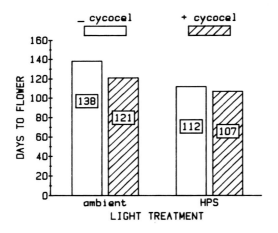

Fig. 7. HPS supplemental lighting reduces length of time to flowering and applying Cycocel reduces it even more. Data from (4).

Supplemental Carbon Dioxide: Carbon dioxide (CO_2) is an essential raw material for the process of photosynthesis. It is present in the atmosphere at a concentration of approximately 330 ppm. On days when ventilation is not required because of low outside temperatures the concentration of CO_2 can fall below 330 ppm (Fig. 8). This often occurs during the winter causing photosynthesis and, therefore, plant growth to be reduced. Even with open ventilators, CO_2 concentrations often stay below that of outside air.

Geraniums are similar to other crops in that they grow more rapidly and are of better quality when treated with CO_2 levels higher than that of normal air.

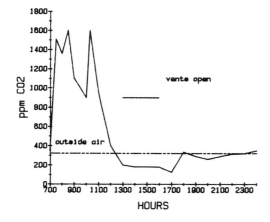

Fig. 8. CO₂ levels often fall below 330 ppm during bright, cold days. Vents opening at 1700 hours barely elevated CO₂ back to normal levels. Adapted from (35).

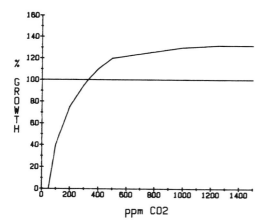

Fig. 9. Growth rates are correlated with CO₂ levels. Adapted from (48).

Growth rates of most plants can be correlated to CO_2 concentration (Fig. 9). As CO_2 levels fall below 330 ppm, the growth is less and as CO_2 is increased above 330 ppm, growth rate increases.

The most efficient level of CO_2 supplementation is usually around 800–1000 ppm. Plants are most sensitive to CO_2 enrichment when they are young. This is similar to the plant response with supplemental light. "Fertilizing" with CO_2 during the entire life of the plant will give the greatest increase in growth; however, the most economically important time for growth increase is the first 4–6 weeks of development. Supplemental CO_2 also results in earlier maturation of geraniums (44), and thus earlier flowering.

Supplemental CO_2 is best utilized by plants grown at higher temperatures and high light levels (either naturally high or supplemental light). This is simply because as light and temperatures are increased, the plant's ability to use CO_2 is also increased. As we become more intensive in our growing methods, the combination of supplemental light, supplemental CO_2, higher temperatures and greater fertility will likely become commonplace in the growing of geraniums (See section on future).

TEMPERATURE

Temperature also influences the length of time to flowering. Both low temperatures (below 10°C, 50°F) and high temperatures (above 32°C, 90°F) slow growth and decrease flowering time due to reduced photosynthetic activity (12) and metabolic processes. Table 7 shows how night temperatures influence flowering time .

Table 7.	The influence of night temperature on length of time to flowering. Data from (36).	

Night Temperature		Days to flower
°F	°C	
50	10	120
60	16	110
70	21	105

Night temperatures of 13–14°C (60–62°F) and day temperatures of approx 21.5°C (70°F) have proved successful in producing high quality hybrid geraniums. If the crop is on schedule, then night temperatures should be reduced to 50–55°F (10–12°C) when the first flowers open.

White and Warrington (51) found significant differences in growth between 10°C (50°F) and 15°C (60°F) night temperatures when day temperatures were 21°C (70°F). They also tested split night temperatures and found there were no differences when the night was split between 15 to 10°C (warm → cool) or 10 to 15°C (cool → warm) (See Table 8).

Table 8.	The effect of split night temperatures on 'Red Elite' geranium.			
Day Temp (°C)	Night Temp (°C)	Leaf Area (cm2)	Total Dry Weight (g)	Time from 6th leaf to visible bud (days)
21	15	1175 a^z	4.9 a	31 b
21	10 to 15	913 ab	3.8 b	30 b
21	15 to 10	890 b	3.4 b	32 ab
21	10	740 b	2.7 c	34 a

^{z}Values within a column followed by the same letters are not significantly different at the 5% level using Tukey's HSD test. Adapted from (51).

Temperature plays a key role in scheduling hybrid geraniums. The time between the flower buds being just visible (>0.5 cm) and their opening to full flower is approximately 25–30 days for most cultivars under normal greenhouse conditions. This stage is directly influenced by temperature and can be speeded up, slowed down or brought to a virtual stand-still by the judicious use of heat so the grower can raise or lower night temperatures to help him meet his sales date. During the spring, every 1°decrease in night temperature, resulted in a 1 day delay between visible bud and flower (Fig. 10). As temperatures warm up during late spring, night temperature changes have less effect due to warmer day temperatures and increased light.

As temperature and light differ in different geographical locales the same cultivars grown in different areas will display different growth and flower habits.

16

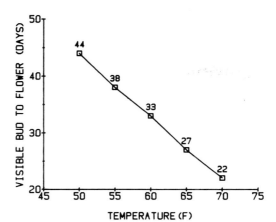

Fig. 10. Increasing the night temperature reduces the time between visible bud and open flower. Adapted from (30).

Table 9 lists some cultivars which were tested at Michigan State University (43°N) and the University of Georgia (33°N) (9).

Table 9.	**Pack trials for geraniums grown in Georgia (UGA) and in Michigan (MSU).**							
Cultivar	Source[1]	Total Height		Flower Diam.		Infl. Diam.		Days to Flower	
		UGA	MSU	UGA	MSU	UGA	MSU	UGA	MSU
White									
White Orbit	GS	24.9	22.3	3.6	4.1	7.8	8.1	96	105
Sprinter White	GS	31.4	27.0	4.2	4.6	9.2	9.7	95	104
Ice Queen	SG	34.5	25.3	3.8	4.3	9.9	9.5	92	106
Salmon-Pink									
Sprinter Salmon	SG	30.5	24.7	4.3	4.5	8.2	9.6	94	110
Cherie 80	SG	28.7	22.7	3.5	3.9	8.7	9.6	99	115
Coral Orbit.	GS	22.6	20.0	3.6	3.8	8.2	8.0	94	106
Appleblossom Orbit	GS	29.4	24.3	3.7	4.1	9.7	9.7	96	106
Red									
Red Elite	GS	21.4	21.7	3.8	4.5	8.2	9.2	92	102
Red Orbit	GS	27.1	26.0	3.4	4.1	8.5	9.7	96	107
Scarlet Orbit	GS	26.4	27.7	3.7	4.0	8.7	8.3	95	102
Ringo Scarlet	SG	30.6	21.0	3.8	4.6	8.3	8.2	93	103
Sooner Red	PA	27.5	20.7	3.6	4.0	9.7	8.3	96	104
Sprinter Scarlet	GS	31.6	23.7	4.2	4.3	8.5	8.9	98	105
Violet									
Merlin	GS	30.1	28.3	3.4	4.2	9.3	10.1	97	108

[1]GS = Goldsmith Seed, Gilroy CA; PA = Pan-American Seed, Chicago, IL; SG = Sluis & Groot Seed, Netherlands.

WARM TEMPERATURE AND HIGH LIGHT FORCING

Some interesting research has shown that geraniums can be forced very quickly by exposing young plants to 85°–90°F (29–32°C) temperatures and high intensity lighting for a short time. For example, 10 day old 'Sooner Red' geraniums

17

treated for 28 days at approximately 2000 f.c. and 90°F, flowered in 65 days from sowing (20). Further work by Bethke (22) showed that plants must be at least 24–30 days old before high temperature and light treatments are effective.

Although these results are only experimental, plug growers could theoretically pretreat plantlets in the plug trays to induce flower initiation prior to shipping. Large numbers of plants may be treated at one time so the cost/plant of such treatments is minimal. Treatments like this have the potential of revolutionizing hybrid geranium culture. There is much more to be learned about high light/heat treatment and this is an area of enormous potential benefit.

Both computer control of greenhouse environment and their use in plant modelling may rewrite environmental recommendations (see section on future). Growers who have monitored leaf temperatures with more sensitive instruments report that geraniums can be grown at different temperatures than thought heretofore. Could it be that because our thermometers and thermostats have been measuring air temperature only, we have been throwing away energy we could have conserved? More information on this interesting development will no doubt appear soon.

COLD CULTURE OF HYBRID GERANIUM

As energy costs have risen, so have the number of crops being grown at cool temperatures. Production of "overwintered" geraniums was a common practice in Victorian times. It was not until the era of steam heating of glasshouses that many crops were grown at "warm" temperatures. In an effort to reduce energy consumption, cold culture of geraniums has been re-evaluated. An outline of the procedure is given below.

Sowing date: This will depend on the amount of active growth achievable before winter temperatures arrive. For Northern U.S. growers a late August–September sowing for March flowering and an October sowing for May flowering is recommended.

Transplant as previously recommended and place in final packs (i.e. 18/flat) or to save space, to smaller pack size (i.e. 32/tray) for the winter season. Grow on at a 60–62°F night temperature and use "free" heat during the day for approximately 2–3 weeks or until night temperatures fall into the 30's and low 40's (0–8°C). Gradually reduce night temperatures to 40–45°F (5–7°C) and "harden-off" the plants. Day thermostats should be set for the same temperature. Ventilate to reduce humidity as fungal diseases are more common at cool temperatures.

The secret is to keep the plants in this cool, hardened condition until you are ready to start them growing once again. Always keep the plants on the dry side but if a spell of warm, bright weather occurs, give a small amount of water in the morning. If a heat wave decides to visit, do not allow it to force the plants into active growth. Ventilate to reduce temperature as much as possible. Do not encourage soft growth because it will be easily damaged when cool weather returns (which it will!).

There are a number of methods for flowering. The first is simply to allow the weather to warm up and revive the plants. This method is the riskiest as one never knows what Mother Nature has in store. A late April–May flowering should be expected with this method. Remember to water modestly as it will take time for the soil temperature to rise to a suitable point for good root growth. Sloppy watering will result in root rot.

The second method is to hold the plants cold and dry until February–March, then raise the night temperature to 60°F to force for April–May. These warm temperatures will be required for only about 5–8 weeks compared to the 14–16 weeks used with a standard January sowing.

The third method is to force early March–April flowering by raising the temperatures in late January–February. Heating costs will be higher than for method 2 but prices may be better.

Advantages of this method are better shipping quality, better zonation and more compact growth. Disadvantages are the possibility of Botyrtis, Pythium and Phytophthera infections, damage to soft growth if unexpected warm weather arrives and more difficulty in precise scheduling. Cultivars which are compact and not vigorous should be used (use of Cycocel during cold temperatures should be avoided). Good results have been obtained in England using this method with 'Video' and 'Gala' cultivars (1).

The third method would likely work better in the Northern U.S. and Canada where winter temperatures are constantly below 40°F for a long period of time. In the Southeastern U.S., winter temperatures are extremely variable making it difficult to maintain "suspended" growth. This is particularly true for glasshouses in the South. Work at the University of Georgia with cold culture of 3 cultivars ('Amaretto', 'Bright Eyes', 'Gremlin Peach Blossom') resulted in a minimum time of 8 weeks in the warm greenhouse for 'Amaretto' when sown Nov. 17, placed in cool house on Nov. 29 and removed to a warm house Mar. 22. Flowering occurred on May 17 (Table 10). Certainly additional work is called for around the country to further evaluate this method.

Table 10. **In the cold culture method, the total time for flowering is much longer than the normal method (60–62°F, 15–16°C NT) but the length of time in the heated greenhouse is significantly less. Unpublished data, University of Georgia.**

Growing method	Total growing time (wks)	Weeks in cool greenhouse	Weeks in warm greenhouse
Cold culture	22	12	10
	24	16	8
	25	19	6
Normal	14.5	—	14.5

Physiological Disorders

Physiological disorders are those which are not insect or disease related. They are usually the direct result of poor environmental control but may be, as in the case of petal shattering, part of the "nature of the beast".

PETAL SHATTER

When growers talk of the physical differences between seed-propagated and vegetatively-propagated geraniums, two of the most common concerns mentioned are that the flowers of the seed geraniums are mostly single and that the petals tend to shatter.

Petal shatter is related to ethylene, even small amounts of ethylene induce petals to fall off (8,10). Ethylene concentration in a greenhouse can be greatly reduced by grooming plants well and maintaining a clean growing area. Grooming plants means removing dead leaves, sweeping away petals which have fallen and generally eliminating decomposing plant tissue which generates that miserable gas.

The most important environmental factor in reducing shatter is lowering temperatures. As plants start to flower, low temperatures are desirable anyway for maintenance and better shelf life. Lowering temperatures from 70°F (21°C) to 40°F (4.4°C) results in a 30% decrease in petal shatter (Figure 11).

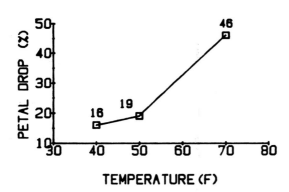

Fig. 11. Petal drop is reduced when temperatures are reduced. Adapted from (10).

Much work has been reported on spraying silver thio-sulphate (STS) to prevent geranium petal shatter. A great number of compounds have been tested to reduce shattering but STS significantly reduced shattering even under warm greenhouse conditions. Subsequently, other research has proved that STS reduces petal drop in a variety of crops. However, STS must not be looked upon as a panacea for petal shatter. First, it is not registered for use in the greenhouse and it is doubtful it will be in the near future. Secondly, there is a potential health hazard to people applying the material due to its extremely rapid and efficient absorption into the skin. Proper clothing must be worn when working with STS. Thirdly, reports have shown that STS increases the sensitivity of geraniums to some of the water molds, particularly Pythium. That is, if Pythium is not present then no problems occur, however, if Pythium is present the plants are more sensitive to it when STS is applied (32). Although there are dangers involved with the use of STS, many growers use it and will continue to do so. An application of fungicide should be faithfully applied 2–3 days prior to its use. Companies presently manufacturing compounds designed to reduce shatter in pot and cut flowers include Floralife Inc., Hinsdale, Ill. and Biendien Naarden of Holland (American distributer, Maroma Inc., Miami, Fla.). However, if one wishes to purchase the necessary chemicals then Table 11 gives an accepted mixing procedure for STS (34).

Table 11. Method for mixing enough STS solution for 1000 geraniums.
Adapted from (34).

1. Dissolve 0.42 g of silver nitrate ($AgNO_3$) in ½ liter of water.

2. Dissolve 2.48 g of sodium thiosulphate ($Na_2S_2O_3$) in ½ liter of water in a separate container.

3. Add all of the silver nitrate solution to the sodium thiosulphate solution while stirring the mixture. If in doubt about which way to mix the solutions, mix them together quickly and the resulting STS solution should be effective.

4. Dilute the resulting STS solution by adding 9 liters of ordinary tap water to give a total volume of 10 liters. In most cases ordinary tap water will be acceptable. However, there may be unusual circumstances unknown to us where compounds in the water cause precipitation of the silver. The presence of brown discoloration can generally be used as a visual check to indicate an ineffective solution.

5. Spray approximately 10 ml per plant. The 10 liters (2.6 U.S. gallons) will cover 1000 plants.

6. Plants can be treated anytime after flower buds are visible but before first florets are open.

7. The dilute STS solution should be used as soon as possible after mixing. If necessary, the STS solution can be safely stored in a refrigerator for up to one month. For longer periods of storage, the silver nitrate and sodium thiosulphate solutions should be stored separately. The silver nitrate solution must be stored out of the light or in a dark glass bottle.

Note: All operations should be carried out using glass or plastic containers. Metal sprayers may be used if the STS is sprayed immediately. Storage in metal containers will result in inactivation of the STS solution.

TEMPERATURE EXTREMES

Extremes of temperatures (above 90°F, below 50°F) cause morphological changes in leaves and flowers. High temperatures result from inefficient venting and cooling equipment or when air flow is poorly distributed. For summer flowering crops, shade on the roof or shade curtains is needed in most parts of the country when light intensities get higher. Cool temperatures may be caused by drafts from broken sash, doors, plastic, or poorly sealed cool pads.

High temperatures cause chlorophyll breakdown in the exposed leaves which often appear chlorotic after a period of 2–4 days (Fig. 12, left, back cover). High temperatures during the seedling stage will cause the seedling to stretch and if sustained, the plants will likely die.

Low temperatures over a long period of time result in clubby stems, large leaves and deformed flowers. The leaves are often highly zoned and flowering time is significantly delayed (Fig. 12, right, back cover). Root organisms, such as *Thielaviopsis,* can also proliferate in cool soil temperatures when temperatures are reduced for "holding" plants.

SOLUBLE SALTS

The measurement of soluble salts is one of the easiest and most important tests a grower can make. Although media can be sent to various laboratories for testing of soluble salts, the purchase of an inexpensive solubridge allows rough monitoring of the fertilization regime. If soluble salts are very low (see Table 12), too little fertilizer is being applied and it should be increased. On the other hand, if soluble salts are too high, damage to the roots will result in marginal leaf burning. Immediately double leach if the reading is high. Leach all pots until water drips out the bottom and repeat 2 hours later. Do not apply any further fertilizer until the soluble salts reading is in the desirable range.

Table 12.	Interpretation of soluble salt readings for hybrid geranium. Adapted from (39).	
Soluble Salt Reading (mmhos (mhos $\times 10^3$))		
1:2 Soil:water	1:5 Soil:water	Interpretation
0.1 to 0.40	0.08 to 0.30	Satisfactory for seed germination, too low for growing on.
0.40 to 1.80	0.30 to 0.80	Satisfactory range for most established plants, may be too high for seedlings.
1.80 to 2.25	0.80 to 1.00	Slightly higher than desired.
2.25 to 3.40	1.00 to 1.50	Plants usually stunted, slow growth.
3.40 +	1.50 +	Severe dwarfing, death will result.

Hand held meters for instant soluble salt readings are available at nominal cost from greenhouse supply dealers and usually come with easy to understand instructions and interpretations. It takes no time at all to make this test—a small price to pay for the difference between crop success or failure.

One of the most common causes of soluble salt problems is failure to calibrate the fertilizer injector. Injectors are notorious for losing calibration. A simple procedure for calibrating the injector, as well as for monitoring fertilizer concentrations is to prepare stock solutions of various concentrations of the fertilizer or fertilizer mixtures used and measure conductivity of each on the meter. Then check the conductivity of the solution from the hose to see if it matches the conductivity of your desired stock solution. (Fig. 13)

Many chemical supply firms have prepared excellent samples of the conductivity ratings of the various fertilizers they manufacture and can supply additional information on this technique.

Some greenhouse supply firms are now selling injectors with a built-in solubridge. The diluted solution passes through the solubridge and if it is not within the desirable range, amendments to the solution automatically raise or lower the conductivity of the solution. These excellent instruments will be commonplace in the near future.

22

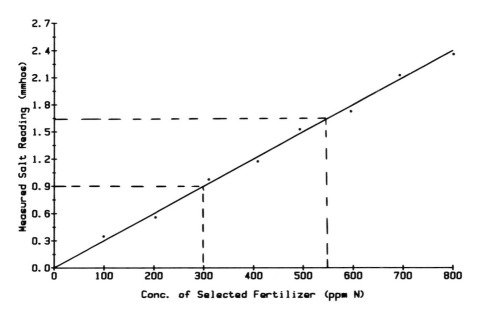

Fig. 13. A straight line should result when various concentrations of fertilizer are read on a solubridge (dots show actual readings). In this example, if a 300 ppm N solution of the same fertilizer was being delivered, the solution should read 0.9 mmhos; for 550 ppm N, the reading should be 1.75 mmhos.

PLANT BLINDNESS

Plant blindness (Fig. 14, back cover) reached epidemic proportions with some bedding plants in England during the 1970s (21). Geranium growers in this country have also experienced this problem. A 2–5% level of blindness is common but at times 15% of the crop may be affected. Distorted, thick, fleshy and brittle leaves show up at the 1st or 2nd true leaf stage and the growing tip of the plant degenerates. Although some plantlets recover to flower normally, most must be discarded. If left unchecked those that do not die will yield only a multistemmed, small-leaved, small-flowered plant which matures slowly.

Whether the problem is caused by an ethylene-induced auxin degradation (26), by nutrition imbalance (21) or other causes remains to be determined. If an emerging seedling is placed under fluctuating temperatures too soon, that is, if seedlings are removed from the germination room before cotyledons have straightened out, then blindness appears to be more serious. Overwatering of the seedlings in the first two weeks can cause plant blindness as well. White cultivars are more susceptible to this problem while it is seldom seen in most of the reds. Salmon cultivars seem to be in-between.

In North America the problem arises more often during periods of low light intensity and more commonly in particular cultivars. Part of the problem may be related to seed treatment and vigor. Plant blindness is one of those problems that is serious enough to notice but it is not regarded to be serious enough to work on by most growers and researchers. This is a grave misconception.

23

Timing and Scheduling

Bedding plant growers are asked each year about the problems they experienced the previous year. The leading internal problem is invariably the timing and scheduling of crops (49). Timing the flowering of hybrid geraniums for a designated week is demanding at best and virtually impossible without advance planning.

Unfortunately, we cannot control weather so if the fall and winter is particularly dark and cold, then plants will be late no matter how much planning has been done. Similarly, if we have warm, bright weather longer than usual, crops will likely be early. Nor can anything be done about the weather during the weekends we are trying to ship our plants. If rain decides to hover about causing our retailers to say "wait", we often must wait and do our best to maintain the plants in top condition. Enough said about the weather!

Flowering times have been presented for many cultivars of geranium over the years in Bedding Plant Inc. *News* (6,7,9,13,31), as well as publications from universities and seed suppliers. Within limits, they are excellent sources of reference but the key to your scheduling is keeping good records so that the effect of local weather can be minimized. That is, records kept over a period of years permit you to anticipate plant response and manage in the context of your own environment.

Simple records, such as shown in Fig. 15, will provide the information necessary to monitor scheduling from year to year. Notice that the visible bud date (bud about the size of a small pea) is included. I believe this to be one of the most important pieces of information you have. Remember that once the flower bud is visible, acceleration or retardation of flowering can be directly controlled by regulating temperature. Since it takes approximately 25–30 days from visible flower bud to open flower at 62°F night, 70°F days for most cultivars, you have that time to fine tune your geranium crop.

SCHEDULING FLOWERING WITH FLAT CULTURE

Let us look at a typical schedule for 'Ringo' geranium and assume a market date of May 1 (Week 18). Based on data taken from issues of Bedding Plant Inc. *News* (9), this cultivar flowers in approximately 93 days in Athens, Georgia and 103 days in East Lansing, MI (See Table 9). The milder climate, fewer clouds and longer days during the winter in Georgia result in reduced time to flower compared to the cooler climate, cloudy winters and shorter days in East Lansing, MI. Similarly 'Ringo Red' grown in Oshawa, Ont., Canada required 108 days to flower sown in January and 114 days when sown in late December. If we use the Michigan data and work backwards 103 days from May 1 you will see that sowing should occur on Week 3. However, if you wish to give yourself an extra 7 days cushion, your actual sow-week will be Week 2.

Transplanting can be scheduled 10–14 days later and Cycocel (if necessary) will be applied week 7 and week 8. Flower buds should be visible approximately 3–4 weeks prior to May 1 (Week 14–15). At that time temperature may be altered to bring your crop in on time (Fig. 15).

Figure 15.

	Sow	Transplant	Growth Regulator	Growth Regulator	Visible Bud	Color in Bud	Market
Schedule for 'Ringo Scarlet' based on environment in East Lansing, MI (9).							
Calendar Week	2	4	7	8	14	16	18
Temp °C	23–26	21D 18N			Alter as necessary	21D 12D	
°F	75–80	70D 62N				70D 55D	
Light	Low	Medium High					
Nitrogen (ppm)	0	150	300			150	0

This schedule assumes good natural day-light, a constant liquid feeding program and no insect or disease problems. Most growers have staggered sales dates so planned production becomes more important due to seasonal changes of light intensity, photoperiod and temperature.

As a general rule, however, the later the sowing date, the faster the crop time. This is because as sowing is delayed, more of the growing occurs during late spring when days are longer, light is stronger and temperatures are warmer. Figure 16 shows graphically how crop time is decreased as sowing is delayed. Records such as these are not difficult to compile and make the job of scheduling so much easier.

PELARGONIUM x HORTORUM

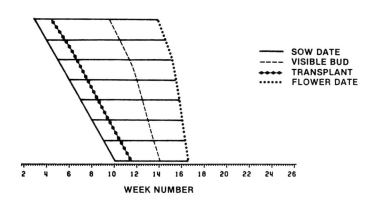

Fig. 16. Long term scheduling of hybrid geraniums. To use the schedule, select the market week on the horizontal axis, and draw a perpendicular line until it hits the "flower date" line. Then draw a line across from that point which cuts across the "visible bud" line, "transplant" and "sowing" lines. Now draw straight lines from those intersections down to the horizontal "week number" axis and read the week for sowing, transplanting and expected visible bud in order to have flowers for the market week selected. Data from (14).

SCHEDULING FLOWERING WITH PLUG CULTURE

Plugs will be grown by the specialist propagator for a certain number of weeks prior to your receiving them. It is essential to know how old the plugs are when they arrive. Some plug producers provide pre-budded material which may only require 2–4 weeks to flower while 40 day old plugs will require 6–9 weeks on the bench.

When finishing geraniums from "plugs", the plants should be transplanted to their final container as soon as possible. If this is not possible, they must not remain in the "plugs" for more than 2 weeks. Leaving the plugs in the plug tray too long results in flowering plants which are both poorly branched and too small for their final container. If they must remain in the plugs, fungicide, fertilizer and growth regulators should be applied.

Scheduling bench space: If you are sowing seed for plug production, remember that your production time is the combination of germination time plus the growing-on time in the plug and growing-on time in the final container. If you are using a 12½" × 20¼", 273 plug tray, each tray will occupy 1.76 ft² while the standard 11" × 22" bedding plant tray will take up 1.68 ft² of growing space. Hybrid geraniums in plugs require approximately 2–3 weeks in the germination area prior to being moved out to the production area. Another 3–4 weeks is customary for growing-on prior to being transplanted to the bedding trays for an additional 9–11 weeks of production (30).

If plugs are not used, then more space will be required due to the longer period of time in the final container.

To demonstrate, let us take an example where 10,000 plants are required. These are sown in approximately 37 plug trays (273 plants/tray) taking up 66 ft² or 100 seed flats (100 plants/tray) taking up 168 ft². The plug trays are moved out in 3 weeks and occupy 66 ft² in the production area for an additional 3 weeks (30). They are then transplanted into 555 bedding flats (18 plants/flat) taking up 933 ft² for 8 weeks. The seed flats are transplanted after 3 weeks, so the main difference in bench space is that the 555 trays occupy the 933 ft³ for 3 weeks longer (assuming an equal 14 week crop time for both systems). Table 13 summarizes the above example.

Table 13.	Crop: Hybrid Geranium					
	Number of plants: 10,000					
	Plug System			*Flat System*		
	Amount	*Space*	*Time*	*Amount*	*Space*	*Time*
Sow	37 plug flats	66 ft²	3 wks	100 seed flat	168 ft²	3 wks
Grow on	37 plug flats	66 ft²	3 wks			
Production	555 flats	933 ft²	8 wks	555 flats	933 ft²	11 wks

Using standard 12½" × 20¼", 273 cavity plug tray and 11" × 22", 18 cavity bedding plant flat.

This example is for northern U.S. production so the times differ further South. I have not inserted the week numbers for sowing, transplanting etc. as that obviously is a decision based on market dates. Some growers have experienced reduced time in final containers when using transplanted plugs. In this case, 8 weeks of production time may become 7 weeks so even more space is saved.

There have been some significant contributions to effective utilization of greenhouse space through the use of linear programming models in this country and in Europe. The models utilize inputs such as number and size of trays or pots needed at a given time, prices for materials etc. These models predict spacing plans designed to fit within specified greenhouse limits. For more details on this important subject, see the sources listed in the bibliography (23,43,50).

POSTPRODUCTION CARE

Retail area: Although the plants may spend up to 16 weeks in the greenhouse, it is the period in the retail store which often dictates whether the plants sell or not. All too often those 16 weeks of tender love and care are wasted due to poor maintenance of the plants in the sales area. Despite the many differences in retail operations and in the people who sell plants there are some rules which must be adhered to if plants are to be maintained in top selling condition (See Table 14).

Table 14.   **Recommendations for retail care of plants.**

Fast turnover is no excuse for sloppy management. Not all material sells at the same rate, nor are all times equally busy. It only takes a few poor looking plants to make the whole area appear shoddy.

Shade all bedding plants from the sun. Both sun-tolerant plants such as geranium or petunia and shade-tolerant plants such as begonia and impatiens must be shaded to slow down deterioration. Approximately 60—80% shade is best for most plants, with 50% being the minimum. Shade may be created by shade cloth over lath, polyethylene, shade trees, etc.

Ventilation is very important to reduce temperature and buildup of pollutants. When displaying outdoors, keep plants from "hugging" large obstructions such as buildings. When buildings shade facilities, be sure that adequate ventilation is provided. If displaying in a greenhouse, turn on fans for plants and people. Circulation of air is nil when plants are packed in tiers or crushed together. This may save space, but the plants deteriorate rapidly.

Raise plants off the ground whenever possible. All too often, plants on the pavement, sidewalk or in the greenhouse sit in a small puddle of water. If the day is hot, the plants are baked. Air and water movement are greatly increased when plants are displayed on a raised bench.

Grooming of the plants (removing dead flowers, dead or dying leaves, etc.) is essential to prevent ethylene gas from being produced in excessive amounts. Grooming also adds to the overall cleanliness of the area. Sales areas must be clean and free of weeds, disease and insects. Allowing trash and half-dead plants to remain is simply asking for a disaster.

People are the most important aspect of the retail sales area. The jobs of grooming, watering, labeling, etc., must be delegated to responsible people. Plants are not furniture and cannot be ignored. If our survey is any indication, plants aren't the problem; people are.

Advertising, good display features, promotions and friendly, informative sales people all contribute to fast turnover. Duane Thompson of Omaha, Nebraska, one of the leading retailers in the floricultural industry, believes that point of purchase sales support must be very strong. He outlines some excellent ideas on retailing, promotion and advertising, all of which helps to have his bedding plants leave the store in as good a condition as when they arrived (46,47).

Greenhouse: Growers can also markedly influence post-production life by carefully choosing the proper container size, a medium which has proper water holding properties, restricting fertilizer and water and lowering night temperature prior to shipping. All of these practices contribute to the longevity of the plants once they leave the greenhouse (16). These and other ways growers can help make the plant last longer are shown in Table 15.

Table 15.	Production procedures to increase shelf life.		
State of Growth	Procedure	Significant (S) or Minor (M) Benefit or Variable (?)	Comment
At transplant	Wetting agent	M	Plants do not require water as often so dry out less readily
At transplant	Use as large a container as possible	S	Plants dry out less readily thus less water stress
Young plants	Use growth regulators where applicable	M→S	Reduce leaf area thus reduce water loss in sales area
Finishing (final 1–2 wks)	Reduce water frequency over the last few weeks	S	Acclimates plants for impending water stress
	Reduce fertilizer frequency and/or concentration	S	Excess nutrients in soil will cause plants stretch as well as a potential salt problem
	Lower greenhouse temperature prior to shipping	S	Helps plants harden off and cope with stress better
	Use anti-transpirants prior to shipping	M(?)	If plants are under minor stress (i.e. warm temp.) may be helpful. If plants under high stress, of little use.

GARDEN PERFORMANCE

Although this manual is intended for commercial growers of geraniums, it is critical that we consider garden performance of our plants as well. After all, it is the garden performance which has made geraniums popular, not how pretty they look in the greenhouse. We'd better all give prime attention to the amateur gardener who puts the bread on our table. Mr. and Ms. Anonymous Gardener will vent their frustration and unhappiness over unsuitable plants with their pocket books. We must grow suitable cultivars—those which we would select to plant at our homes, not those which look good in a pack in the greenhouse.

Flower gardening is the number 1 outdoor leisure activity in the United States. It is obviously impossible for growers to plant out and evaluate all the new cultivars of hybrid geraniums which appear each year, let alone the hundreds of cultivars already on the market. Not all cultivars look in gardens the way they look in catalogues. Some cultivars simply are not good for a particular region (North, South, etc.) due to heat, cold, rain, drought or whatever conditions are commonly found during the growing season. Although some cultivars have great pack performance, they should not be sold to gardeners in unsuitable areas.

It is likely that there are trial gardens in your region where some of the newer cultivars of geraniums are compared to standard cultivars. Trial gardens may be part of a seed company's trials, a service of public gardens or there may be a University in your area with trial grounds. Many people look at these; you might as well be one of them. Many of these gardens publish extensive results on all cultivars in their trials and are a good source of information.

In addition, there are approximately 30 All America Trial Gardens in North America which evaluate geraniums (and other bedding plants), some of which will not be marketed for another 2–3 years. These gardens are spread across this continent and with its European counterpart, Fluorselect, many locales are being used. If the cultivar wins an All America award, one can be reasonably sure of satisfactory garden performance regardless of the region; however, experience and judgment based on your local circumstance are invaluable. To obtain the addresses of All America Trial Gardens write All America Selections, 628 Executive Dr., Willowbrook, IL 60526.

Seeing plants in their garden setting is a much more realistic measure of their performance than simply admiring their pack performance. Trials certainly do not answer all the questions but they will help you and your customers stay abreast of what's new in a constantly changing marketplace.

Pests and Diseases

One of the main advantages of seed propagated geraniums is their relative freedom from fungal, bacterial or viral diseases. However, there still are a number of pests and diseases which prove troublesome. The best defense for pest and disease problems is good sanitation. It is seldom possible to cure a plant once a disease organism has caused visible plant stress, and leaves which have been damaged from insect infestation will not revert back to normal.

A word about chemical control: In this section, I mention the use of certain chemicals for control of insects and diseases. Throughout I will be repeating the axiom that "good sanitation is good business" in one form or another. Although most growers use chemicals for pest control, those who *practice* good sanitation programs are those who use chemicals only minimally. Chemicals do not replace good growing practices.

There are many chemicals available and the only thing which is certain today is that they will be different tomorrow. I have mentioned only a handful I thought to be appropriate but there are many more which may be effective for your particular problem. Many of the new chemicals soon to be available may provide better control than those mentioned in this section. It is important to check with your university extension pathologist or entomologist and chemical distributors for currently available materials and application updates. There are also a number of excellent pertinent publications on pest control available. A fine overview of pests and their control may be found in Cornell Recommendations for Commercial Floriculture Crops, Part II, Pest Control (3) available from Cornell University Distribution Center Research Park, Ithaca, NY 14850 at a nominal cost.

PESTS

There are amazingly few pests which trouble hybrid geraniums. Although we associate fuchsia with whiteflies, hibiscus with aphids and roses with spider mite, geraniums are not attacked in great swarms by pests. There are, however, a few insects worth mentioning.

Aphids: These soft-bodied adults and nymphs are found in groups on leaves and stems. They not only injure the plant by sucking the sap with their piercing mouthparts but also are natural carriers of viruses from other plants in the greenhouse. The honeydew they excrete is very unsightly and is a medium for growth of black, sooty mold. Control with aphicides such as Metasystox-R EC (1½ pts/100 gal, 180 ml/100 l), Malathion 25% WP (1 lb/100 gal, 100 g/100 l), and Piramor.

Caterpillars: This diverse group of insects are immature forms of butterflies and moths. Although the adults do not feed, the larvae more than make up for it by chewing off large pieces of leaves or eating the meristematic region, leaving unsaleable plants in their wake. Many moths are attracted into the greenhouse by lights whence they lay their eggs in the soil. Control of caterpillars on geraniums has been successful with dichlorvos (Vapona, DDVP, 1 lb/100 gal, 100 g/100 l, carbaryl (Sevin, 50% WP, 1 lb/100 gal, 100 g/100 l), trichlorfon (Pylox, 80%

SP, 20–30 oz/100 gal, 150–220 g/100 ml), or *Bacillus thuringiensis* (Dipel, Thuricide).

Fungus Gnats: These greyish black flies lay their eggs in the growing medium. The larvae then start chewing on root hairs and organic matter in the medium. The larvae are white with black heads and may be found in large numbers, particularly if the medium has not been allowed to dry out between waterings. The larvae change to pupae which emerge as adult flies 3–4 days later. The flies are disturbed easily so are quite noticeable before serious injury occurs. Malathion and Diazinon drenched according to the label gives adequate control.

The following insecticides have caused plant injury in geraniums: acephate (EC), demeton, dimethoate, nicotine sulfate (smoke), morestan, oxydemetonmethyl, and endosulfan (3).

DISEASES

Many of the crippling diseases spread by vegetative cuttings or by contaminated knives are effectively reduced with seed propagation. However, seed propagated geraniums are just as sensitive to disease as vegetatively propagated geranium and diseases of a vegetatively propagated crop may still occur in seed propagated geraniums if the vegetative crop is present in the greenhouse.

Root rots: *Pythium ultimum, Rhizoctonia solani* and *Thielaviopsis basicola* not only cause damping-off of geranium seedlings but also can result in root and basal rot of older plants. All are spread by infested soil or plant tissue. These diseases can be devastating, which is one more very important reason that cleanliness is next to godliness in the greenhouse. Control methods to be considered are disinfection of the mix, benches, tools etc. that come in contact with the plant, use of a light, well-drained soil mix, and supplementary chemical soil drenches to minimize contamination. Chemical control of these organisms includes separate application of Lesan (35% WP) or Terrachlor (75% WP) each at 5 oz per 100 gal (60 g/100 l) or a combination of the two (see Damping Off). Lesan breaks down in the presence of sunlight and must be used immediately upon mixing. Terrachlor will not control water molds (e.g. Pythium) and frequent use may cause some phytotoxicity. Subdue may also be used at ½ g per 100 gal (6 g/100 l).

Damping-Off: Damping-off of seedlings is caused by a host of organisms including species of Rhizoctonia, Pythium and Phytopthora. Movement of these organisms is mainly by mechanical transfer of the organisms in infested soil particles (end of the watering hose, etc.) or infected plant tissues. Effective control is, therefore, easiest by steam sterilization or surface disinfection of seed medium and good sanitation practices. Lesan (35% WP) and Terrachlor (75% WP) at 8 oz and 4 oz per 100 gal (60 g and 30 g/100 l), respectively can be used together as a soil drench for control of Pythium and Rhizoctonia species.

Botrytis spp.: This organism causes gray mold. Many plants are susceptible, especially when they are grown under cool, humid conditions. Since maintaining cooler temperatures is becoming an increasingly common production technique, this disease is likely to become even more prevalent. Symptoms of this disease first appear as stem lesions and leaves become covered with a "gray fur". One of the best means of control is to heat during the late afternoon, then vent out the humid air and replace it with drier air from outside. This fungus generally is found where there is dead or decaying tissue on which it may rest. Once again, strict sanitation including daily grooming of plants is essential. Chemicals used for Botyrtis are Exotherm Termil® smoke and Daconil 2787® (1½ lb/100 gal, 150 ml/100 l).

Others: Diseases which are of concern to growers of vegetatively propagated geraniums such as Bacterial Stem Rot, Leaf Spot and Verticullium Wilt are not nearly as crippling with seed propagated geraniums. It would, of course, be foolish to say that these diseases do not exist at all because they may be lurking on other plants somewhere in the greenhouse and will infect seed propagated geraniums if conditions are right. Therefore, for all diseases follow a strict preventative control program. This means that you must keep foliage dry at all times, check incoming plants carefully and destroy damaged or spent plant tissue such as infected leaves or dried blossoms. Chemical control takes a distant back seat to good cultural control in the prevention of disease.

Costs of Growing Hybrid Geraniums

Hybrid geraniums are like any other crop in the greenhouse and the final selling price must reflect both direct growing (allocated) costs as well as overhead (unallocated) costs. In assigning figures for various costs, I found tremendous variations between growers, especially for overhead expenses. Wholesale prices for a flat of 18 hybrid geraniums ranged from $4.50 to $9.00, and the overall sentiment was to charge what the market would bear. There were especially large discrepancies in assigning cost/ft^2/wk figures. Many growers had a poor idea of these costs so I have presented an example of how to determine cost/ft^2/wk based on growing 3 crops of 20,000, 18 cell flats of geraniums per year based on the model of Strain (45). The greenhouse is in operation 47 weeks of the year.

Table 16.   **Determining cost/sq ft/wk of growing hybrid geraniums from seed (adapted from 45)**

Example: for 34,600 ft^2 of growing space (40,000 ft^2 actual greenhouse space): 3 crops of 20,000 flats (18 plts/flat) each: Average crop time = 14 wks: 42 wks growing time.

Line	Item	Cost from greenhouse records	Less sum of costs directly related to geraniums	Overhead costs
1	Salaries, wages (Owner-manager plus employee wages and benefits)			80,000
2	Seeds, @ 5¢, 85% germ.	63,530	63,530	
3	Containers, @ 55¢/flat	33,000	33,000	
4	Soils, @ 40¢/flat	24,500	24,000	500

Line	Item	Cost from greenhouse records	Less sum of costs directly related to geraniums	Overhead costs
5	Fertilizer	6,300	6,000	300
6	Repairs, maintenance			4,000
7	Other direct e.g. tags	1,200	1,000	200
8	Costs directly assignable to whole crop		127,530	
9	Costs directly assignable to each of the 3 crops		42,510	
10	Utilities (based on growing area)			60,000
11	(Administrative costs, selling costs)			4,500
12	Fixed overhead (Insurance, Depreciation, Rent, Taxes)			22,500
13		Total unallocated		172,200
14	Investment in greenhouse		190,000	
15	Desired rate of return		.20	
16	Unallocated return in investment (14 × 15)			38,000
17		Total unallocated costs (16 + 13)		210,000
18	Net growing area (ft²)			34,600
19	Unallocated costs/ft²/year (17 ÷ 18)			6.06
20	Weeks in operation			47
21	Unallocated cost/ft²/wk (19 ÷ 20)			0.129

Calculating Costs for Empty Space

Line	Item	Value		
22	Net growing area	34,600		
23	Weeks in operation	47		
24	Total time space (22 × 33)	1, 626,200		
25	Growing time space (20,000 × 1.68 ft × 14 wks) × 3	1,411,200		
26	Empty time adjustment (24 ÷ 25)	1.15		
27	Unallocated cost/ft²/wk (26 × 21)	0.148		

After all is said and done, the geraniums are costing approximately 15¢/ft²/week in this greenhouse. Your greenhouse costs may be entirely different, but the concept to determine cost/ft²/wk is the same.

NOW WHAT?

In order to determine what it costs you to produce a flat of hybrid geraniums from seed, an example follows for the same situation. An average of 14 weeks for each crop is used (see Table 17).

Table 17.			Determining costs for a flat of geraniums from seed.						
Direct (allocated) Costs			*Overhead (Unallocated Costs)*						
Col 1	*Col 2*	*Col 3*	*Col 4*	*Col 5*	*Col 6*	*Col 7*	*Col 8*	*Col 9*	
Sum of direct cost	No. of units	Direct cost per unit	Space/ unit (ft^2)	Space/ crop (ft^2)	weeks	Space-weeks	Cost/ft^2/ wk	Indirect cost per unit	
42,510	20,000	2.12	1.68	33,600	14	470,400	0.15	3.53	

	Cost Per Unit	
Col 10	*Col 11*	*Col 12*
Total cost per unit	Percent saleable	Adj cost per unit
5.66	.95	5.95

Direct Costs

Column 1—Sum of costs directly assignable to the crop. Line 9, Table 16.
Column 2—No of units grown each time.
Column 3—Direct cost/flat. Col 1 ÷ by Col 2.

Overhead Costs

Column 4—Space/flat: 1.68 ft^2 per standard flat.
Column 5—Space/crop: Col 4 × Col 2.
Column 6—Weeks to grow crop (incl. germination).
Column 7—Time on bench × space occupied: Col 5 × Col 6.
Column 8—Cost/ft^2/wk: Line 27, Table 16.
Column 9—Unallocated cost/flat: (Col 8 × Col 7) ÷ Col 2.
Column 10—Total cost: Col 9 + Col 3.
Column 11—Percent saleable either projected or from records.
Column 12—Actual cost/flat: Col 10 divided by Col 11.

In the case of this grower, the price which he must change is $5.95 per flat just to break even. To determine the price to charge to make a profit, refer to Table 18. For a desired profit of 15%, multiply the cost ($5.95) by 1.176. The resulting charge per flat is $6.95.

This example is obviously a very simple one based on flats of geraniums only. However, if the grower was producing 4" pots or finishing plugs, etc., the concepts are exactly the same. The bottom line here is that you must know what it costs to produce a crop before you know how much to charge.

Table 18.	Profit factors for achieving desired profits on sales.					
Profit on sales wanted	10%	15%	20%	25%	33⅓%	40%
Profit factor						
Multiply cost by . . . to obtain selling price	1.111	1.176	1.25	1.333	1.50	1.66

FINISHING GERANIUMS FROM PLUGS

In the case of the specialist grower who is only finishing plugs, the types of costs are the same but some of the numbers will be different. Using the above example let us assume a cost of $0.15/ft2/wk for basic greenhouse costs. (Actually because of fewer weeks of operation, basic greenhouse costs are lower in this example.)

The only change in Table 16 is at line 2. Instead of seeds at 5¢ each, 6 wk old plugs are purchased for approximately 11¢ each. The costs directly assignable to each of the 3 crops (line 9, Table 16) is $60,933.00 instead of $42,510.00.

Substituting the figures for plug production in Table 17, we obtain Table 19 for determining costs for a flat of geraniums from plugs.

Table 19. **Determining costs for a flat of geraniums from seed.**

Direct (allocated) Costs			Overhead (Unallocated Costs)					
Col 1	Col 2	Col 3	Col 4	Col 5	Col 6	Col 7	Col 8	Col 9
Sum of direct cost	No. of units	Direct cost per unit	Space/ unit (ft²)	Space/ crop (ft²)	weeks	Space- weeks	Cost/ft²/ wk	Indirect cost per unit
60,933	20,000	3.05	1.68	33,600	8	268,800	0.15	2.02

Cost Per Unit		
Col 10	Col 11	Col 12
Total cost per unit	Percent saleable	Adj cost per unit
5.07	.95	5.33

Direct Costs

Column 1—Sum of costs directly assignable to the crop. Line 9, Table 16.
Column 2—No of units grown each time.
Column 3—Direct cost/flat. Col 1 ÷ by Col 2.

Overhead Costs

Column 4—Space/flat: 1.68 ft² per standard flat.
Column 5—Space/crop: Col 4 × Col 2.
Column 6—Weeks to grow crop.
Column 7—Time on bench × space occupied: Col 5 × Col 6.
Column 8—Cost/ft²/wk: Line 27, Table 16.
Column 9—Unallocated cost/flat: (Col 8 × Col 7) ÷ Col 2.
Column 10—Total cost: Col 9 + Col 3.
Column 11—Percent saleable either projected or from records.
Column 12—Actual cost/flat: Col 10 divided by Col 11.

In this case, a finisher must charge $5.33 to break even. For the same 15% profit the flat should be sold for $6.26. The numbers presented here will probably not reflect the number for your own operation, but the method can be used to calculate figures from your existing records.

Future

Looking into the future is fraught with question marks. Making claims concerning the future of anything is best left to soothsayers, doomsday prophets and weathermen—and their rate of success has not been overwhelming. Looking towards some of the things going on today which will affect hybrid geraniums in the near future involves some risk, but it is an exercise that will perhaps stimulate some thought and ideas. Here are some of mine:

BREEDING

Plant breeders are a wonderful hardy lot. They make hundreds upon hundred of crosses which result in thousands upon thousands of potential cultivars and are ecstatic if one or two are worthy of trial. These are the people who provide the building blocks of our industry. Their tenacity and fortitude is something I deeply admire.

In the near future, we will be seeing more and more non-shattering semidouble and double seed geraniums which will flower in the same length of time as the singles. There also will be continued breakthroughs in ivy-leaf, hanging geraniums from seed. Soon there will be single colors which flower in the same time as the zonal types.

Although we have more than enough colors available, there inevitably will be more reds, cerises, purples, mauves and burgundys. I believe the next new color to arise will be a pure orange flowered geranium. I have seen some experimental cultivars very close to a clear orange and within 2 years orange should be available in hybrid geraniums.

One of the important areas which breeders must always keep in mind is uniformity and viability of the seeds produced by their crosses. As more and more growers rely on automated seeders and plug trays, the cultivars produced commercially will be those with high germination percentages and uniform germination in the tray.

The main thrust of breeding, in my opinion, should be towards fine tuning the form of the plant and particularly the flower head. There is little doubt that more genetic dwarf cultivars will be available in the near future for which growth regulators will not be necessary. However, vigor is a trait we don't want to lose in new geraniums and growth regulators will continue to be employed on many of the popular cultivars. Concerning growth regulators, I think more growers will be using the new generation of chemicals such as "Bonzi" as more data become available. They do the same job as Cycocel at lower concentrations and do not show any marginal burn. At the time of this printing, however, Bonzi is still experimental for hybrid geranium.

I see breeding for flower habit going in two directions. The first is to obtain simultaneous opening of multiple florets on the first inflorescence. Many cultivars open with 1 or 2 florets and give the impression of a very loose inflorescence. Although a full flower head is finally attained when most of the florets open, breeding for simultaneous opening of 6–10 florets would make a far better plant picture in the pot or flat. This breeding is already seen in some of the newer cultivars and I believe this characteristic of "early fullness" will soon become more common.

The second direction is further away, but nevertheless attainable. I think, in a few years, we will see "multiflora geraniums"; that is, cultivars which send up 2 or more inflorescences at the same time. This will be a fantastic breakthrough because rather than a single flower head in color, perhaps half a dozen inflorescences displaying "early fullness" will be present when the crop is sold. The first of these is likely to be at least 2 years away, but it will shake up the industry.

Breeding for earlier flowering will also continue. It is amazing that the norm for the "New Era" series was over 20 weeks a few short years ago and today, 10–12 weeks without extra lights or CO_2 is expected of some of the newer cultivars.

Breeders have indeed worked miracles with the hybrid geranium and the future in this direction is bright.

PRODUCTION AREAS

I believe there will be large changes in the areas we use to grow geraniums (and other crops as well). The future I am about to unfold is more likely to occur in North America than in the floriculturally developed areas of Northern Europe. This is mainly attributable to the tremendous differences in temperature in this country compared with those in Europe where the climate of many countries is modified by oceans. Vast differences in North American environment make plant environments much more difficult to control, and control is the name of the game. All the stacks of research done on geranium in the past has shown certain truths to be self evident.

(1) Geraniums are much more responsive to additional environmental inputs such as supplemental light, CO_2, warm temperatures, etc. when they are young (1–4 wks old) than when they are older.

(2) Control of these inputs is much more difficult in a greenhouse than in growth rooms.

(3) An older geranium can be "finished" in almost any environment, as long as temperatures are somewhat controlled.

Recent innovations in technology and automation have resulted in accurate environmental sensors, as well as accurate and inexpensive means to provide the extra inputs such as high output lights, and new efficient CO_2 generators. Lastly, the computer has entered the world of flower production and completed the chain. This means that a correct environment can now be produced efficiently, and measured efficiently. Much more importantly, the computer can analyze the environment, make decisions based on known plant responses to the environment, and change the environment to optimize growing conditions, as necessary. In the case of hybrid geraniums, optimization of conditions for rapid flowering may be chosen or perhaps plant size may be more important than flowering time so the condition leading to these objectives can be optimized. In the case of vegetatively propagated geraniums, rooting response, high numbers of breaks on stock plants or growth and flowering may be optimized by feedbacks of the environment from sensors to computer and back to the plants. To realize these outcomes growing areas must reflect the technology available. One possibility is shown in Fig. 17. It consists of 3 different growing areas on various levels of sophistication, but perhaps only 2 (delete the 2nd area) would be necessary.

GROWING AREA 1	GROWING AREA 2	GROWING AREA 3
HIGH TECH	HIGH TECH	LOW TECH
"GROWTH CHAMBER"	GREENHOUSE	GREENHOUSE
SUPP. LIGHT, CO_2	CO_2, TEMP. CONTROL	FINISHING AREA
TEMP. CONTROL	COMPUTER DRIVEN	TEMP. CONTROL
COMPUTER DRIVEN		
0-8 LEAF STAGE	8 LEAF STAGE - V B	V B - FLOWER

Fig. 17. A schematic outline for future production of floriculture crops.

There are 2 main arguments (and I have heard many!) against the scenario I have just presented.

The first is cost. It is not yet feasible economically to jump into this approach. I believe that as technology advances, costs of technology will continue to drop. All we have to do is look at the cost of microcomputers today compared with 5 years ago. The physical structures for the high tech areas are not elaborate and cost far less than a glass or acrylic greenhouse.

Secondly, and far more important, is that the computer is only as smart as the data we provide. Today, there simply are not enough data available for high tech production. Research into environmental control of crops continues to provide good information. Further, crop modelling is starting to gain momentum and will provide rich rewards in the very near future. Crop modelling is the prediction of the plant response based on changes in its environment. As more research is done in this field, the closer we will come to teaming technology with information to produce crops based strictly on plant response to the environment.

There will also be many other changes including new innovations in heating systems (better bottom heat perhaps combined with more economical infrared heat), cooling systems and management.

SPECIALIZATIONS

There will be more growers who concentrate on single crops on a year-round basis rather than on a multitude of crops. One of the most important trends is that more growers will grow only part of the crop (e.g. plug producers only, finishers only). This allows growers to be more efficient by concentrating on a particular phase of the plants production. They can do what they know how to do best. Efficiency is so important that increased specialization will definitely be part of the future.

There is, however, a dark side to specialization. The question arises "who dictates the cultivars to be grown?" Breeders base much of their work on market forces which tell them what traits are most desirable. For years, breeders have been developing cultivars which provide good "pack performance" because this is what growers want. The cultivar may have abominable outdoor performance, but if it has good pack performance, it will stand a better chance of being profitable for the seed producer. I believe today there is more of an awareness of outdoor performance (i.e. customer satisfaction) and the most successful cultivars are those which have shown both good pack and field performance. This field performance, however, will again be diminished as more large plug growers dictate cultivar characteristics. They will demand cultivars which respond well in their automated production scheme and will dictate that breeding be done for "plug performance". They will want fewer cultivars and will be big enough in the near future to dictate the direction of breeding. For instance, they may base their orders on cultivars which grow very quickly in the first 4 weeks of growth, or only those which are particularly uniform or those from one series only (e.g. Orbits) which provide the colors they need and are similar in habit and response. There will be a falling out of many cultivars due to specialization and for better or worse, plug growers will determine cultivar availability in the future.

MARKETS

The good news is that I believe there will be increased awareness by the public of floriculture goods. We see today a growing demand for new crops, be they potted or cut flowers, and this demand is sure to increase. We need to make our product more accessible all the while observing the age old tenet of "good quality at a fair price".

The future of floriculture on a world-wide basis is bright. The future of geraniums in general is even brighter. Growers of hybrid geraniums in particular will also enjoy its light.

Appendix

Determining air porosity of a mix: In this example, plugs are used, however, the same procedure may be used for pots. (Adapted from 38)

1. Measure the volume of the plug. Do this by securely taping the holes at the bottom and fill the empty plug with water to the soil line. Mark this line with a pencil. Carefully pour the water from the plug into a measuring cup. Do this with 10 plugs in order to get enough water to measure. Divide the total amount of water obtained by 10 to obtain the "total volume" of the plug.
2. Next, dry the inside of the plug. Do not remove the tape. Fill the plug with dry medium. Pack it as you would when transplanting a seedling. Do this with 10 plugs.
3. Using a measuring cup, wet the medium, and measure the amount of water it takes to thoroughly saturate the medium. When a thin film of free water appears at the soil line—this is, when the medium is water-saturated—stop. (Some media are more difficult to wet than others; dry peats may take a long time to saturate. Water absorption can be hastened by applying hot water.) The total amount of water added indicates the "total porosity"—i.e., what percent of the medium consists of spaces between and within particles. These pore spaces can be occupied by water or air. To obtain percent porosity from the following equation of

$$\% \text{ Total Porosity} \;=\; \frac{\text{Water needed for saturation}}{\text{Volume of the plug}}$$

4. Once the medium has been thoroughly saturated, elevate the plugs above the bottom of a receptacle and remove the tape from the holes. Water will drain from the plugs; allow them to drain until no more water comes out. Measure the amount of water that has collected in the receptacle. This volume of drained water is equivalent to the air space in the drained medium. This provides the percent of air space by:

$$\% \text{ airspace} \;=\; \frac{\text{Volume of drained water}}{\text{Volume of the plug}}$$

The percentage of air space (percent of the total volume of the drained medium that is occupied by air) equals the amount of drained water divided by the total volume of the plug. The difference between the amount of applied and the amount of drained is the *water-holding capacity* of the medium.

$$\text{Water holding capacity} \;=\; \% \text{ porosity} \;-\; \% \text{ airspace}$$

With these figures, a mix can be evaluated to determine percent air space and waterholding capacity. If necessary, adjustments can be made. For example, a medium composed primarily of small particles has small pores and tends to retain more water, and consequently less air, than a medium having large pores. The ratio of the various medium components and particle sizes and shapes can be adjusted to the specific container, plant requirements and irrigation.

References

1. Anonymous. 1982. Low Temperature geraniums. Floranova Ltd. *Technical Bul.* 1, 4 p.
2. Anonymous. 1979. *Cornell recommendations for commercial floriculture crops.* Part I. Cultural practices and production programs. Cornell University, Ithaca, NY.
3. Anonymous. 1979. *Cornell recommendations for commercial floriculture crops.* Part II. Pest control—disease, insects, and weeds. Cornell University, Ithaca, NY.
4. Armitage, A. M., M. J. Tsujita and P. M. Harney. 1978. Effects of Cycocel and high intensity lighting on flowering of seed propagated geraniums. *J. Hort. Sci.* 53:147–149.
5. Armitage, A. M. and M. J. Tsujita. 1979. The effect of supplemental light source, illumination and quantum flux density on the flowering of seed propagated geraniums. *J. Hort Sci.* 54:195–198.
6. Armitage, A. M. A. Eurich and W. H. Carlson. 1979. Hybrid geranium greenhouse pack trials 1979—MSU. *BPI News,* July, 5–6.
7. Armitage, A. M., W. H. Carlson and A. Eurich. 1980. Geranium pack trials—1980—more consistency than ever. *BPI News,* Nov. 4.
8. Armitage, A. M., R. H. Heins, S. Dean and W. H. Carlson. 1980. Factors affecting petal abscission in the seed-propagated geranium. *J. Amer. Hort. Sci.* 105:562–564.
9. Armitage, A. M., C. C. Bethke and W. H. Carlson. 1981. Greenhouse trials of hybrid geranium production. *BPI News,* 12(10):1–2.
10. Armitage, A. M. and W. H. Carlson. 1981. Hybrid geranium shatter. *BPI News,* April 5.
11. Armitage, A. M. and C. Lasco. 1981. Effects of time of transplanting on bedding plant flowering. *BPI News,* Sept. 14.
12. Armitage, M., W. H. Carlson and J. A. Flore. 1981. The effect of temperature and quantum flux density on the morphology, physiology and flowering of hybrid geraniums. *J. Amer. Soc. Hort. Sci.* 106:643–647.
13. Armitage, A. M. 1982. Greenhouse and garden trials of hybrid geraniums—1982. *BPI News* 13(12):4–5.
14. Armitage, A. M. 1983. Influence of sowing date on the scheduling of bedding plants. University of Georgia Agr. Exp. Sta. *Research Bull.* 307:15pp.
15. Armitage, A. M., H. M. Vines, Z. P. Tu and C. C. Black. 1983. Water relations and net photosynthesis in hybrid geranium. *J. Amer Soc. Hort. Sci.* 108: 310–314.

16. Armitage, A. M. 1984. Bedding plant shelf life—where we are and where we are going. *BPI News* 15(1):1–3.
17. Armitage, A. M. 1984. Early fertilization of bedding plant seedlings. Ga. Comm. Flow. *Gro. Notes,* Nov–Dec 9.
18. Armitage, A. M. 1984. Effect of leaf number, leaf position, and node number on flowering time in hybrid geranium. *J. Amer. Soc. Hort. Sci.* 109:233–236.
19. Armitage, A. M. and H. Y. Wetzstein. 1984. Influence of light intensity on flower initiation and differentiation in hybrid geranium. *J. Amer. Soc. Hort. Sci.* 19:114–116.
20. Armitage, A. M. 1980. *Effect of light and temperatures on physiological and morphological responses in hybrid geraniums and marigolds.* Ph.D. Thesis. Michigan State University, pp. 65–70.
21. Armitage, Mary. 1980. Plant blindness. *BPI News,* Nov. 5–6.
22. Bethke, C. 1984. Geranium flowers in 50 days with temperature and light. *BPI News* 15(12):11–13.
23. Busham, C. W. and J. J. Hanan. 1983. Space optimization in greenhouses with linear programming. *Acta Hortic.* 147:45–56.
24. Cairns, E. J. and L. M. Lenz. 1982. F. Hybrid geranium seed—effect of temperature on germination. *BPI News* 13(10):4–5.
25. Cairns, E. J. and L. M. Lenz. 1982. F. Hybrid geranium seed—effects of media, moisture and fungicide in germination. *BPI News* 13(11):2–3.
26. Carlson, W. H. 1980. Plant blindness also problem in U.S. *BPI News,* Nov. 7.
27. Carpenter, W. J. and R. C. Rodriguez. 1971. Earlier flowering of geranium in Carefree Scarlet by high intensity supplemental light treatment. *HortScience* 6:206–207.
28. Craig, R. and D. F. Walker. 1963. The flowering of *Pelargonium hortorum* Bailey seedlings as affected by cumulative solar radiation. *Proc. Amer. Soc. Hort. Sci.* 83:772–776.
29. Eastburn, D. P. 1984. The plug potential. *Flor. Rev.* 174 (4497):26–29.
30. Freeman, R. 1983. Plug scheduling. *BPI News* 14(12):1–3.
31. Hamilton, B. M. and A. M. Armitage. 1983. Hybrid geranium pack and garden performance. *BPI News* 14(11):2–4.
32. Hausbeck, M. K., Stephens, C. T. and R. D. Heins. 1984. STS/Pythium interaction . . . does it exist. *BPI News* 15(4):1–2.
33. Heins, R. D. 1979. Influence of temperature or flower development of geranium 'Sprinter Scarlet' from visible bud to flower. *BPI News.* Dec. 5.
34. Heins, R. D., Fonda, H. N. and A. Cameron. 1984. Mixing and storage of silver theosulphate. *BPI News* 15(7):1–2.
35. Heij, G., and P. J. A. L. de Lint. 1984. Prevailing CO_2 conditions in glasshouses. *Acta Hortic.* 162:93–100.
36. Konjoian, P. S. and H. S. Tayama. 1978. Production schedules for seed geraniums. *Ohio Flor. Ass'n Bul.* 579:1–2.
37. Koranski, D. Personal communication.
38. Laffe, S. R. and D. S. Koranski. 1984. Keeping plugs under control. *Flor. Rev.* 174(4497):34–39.
39. Masterlerz, J. W. 1977. The Greenhouse Environment. *The effect of environmental factors on greenhouse crops.* J. P. Wiley & Sons.
40. Mitwally, A. W., B. E. Struckmeyer and G. E. Beck. 1970. Effect of three soil moisture regimes on the growth and anatomy of Pelargonium hortorum. *J. Amer. Soc. Hort. Sci.* 95:803–808.
41. Miranda, R. M. and W. H. Carlson. 1980. Effect of timing and number of applica-

tions of chlormequat and ancymidal on the growth and flowering of seed geraniums. *J. Amer. Soc. Hort. Sci.* 105:273–277.

42. Reiss, W. C. and H. K. Tayama. 1981. Optimum germinating temperature for seed geraniums. *Ohio Flor. Ass'n Bul.* 616:6.

43. Schumacher, S. and F. C. Weston. 1983. Linear programming—a tool for production planning in glasshouse floriculture. *Acta Hortic.* 147:83–88.

44. Shaw, R. J. and M. N. Rogers. 1964. Interactions between elevated carbon dioxide levels and greenhouse temperatures on the growth of roses, chrysanthemums, carnations, geraniums, snapdragons, and African violets. *Flor. Rev.* 135(3486):23–24, 88–89; 135(3487);21–22, 82; 135(3488):73–74, 95–96; 135(3489):21, 56–60; 135(3491):19, 37–39.

45. Strain, J. R. 1984. How to figure your cost to grow bedding plants. *BPI News* 15(12):5–10.

46. Thompson, D. 1983. Bulb promotion for early fall sales. *BPI News* 14(7):1–2.

47. Thompson, D. 1983. Comparing advertising results. *BPI News* 14(4):1.

48. Van Berkel, N. 1984. CO_2 enrichment in the Netherlands. *Acta Hortic.* 162:197–205.

49. Voight, A. 1984. Weather dampens '83 bedding plant season . . . nice increase slated for '84. *BPI News* 15(4):5–6.

50. Weston, F. C. and S. Schumacher. 1983. Economic interpretation of model results in a glasshouse floriculture environment. *Acta Hortic.* 147:89–98.

51. White, J. W. and I. J. Warrington. 1984. Effects of split night temperatures, light, and chlormequat on growth and carbohydrate status of *Pelargonium* × *hortorum. J. Amer. Soc. Hort. Sci.* 109:458–463.